Recycle, Reuse

by Jim Cordray

Harcourt
SCHOOL PUBLISHERS

Orlando   Austin   New York   San Diego   Toronto   London

Visit *The Learning Site!*
www.harcourtschool.com

Introduction

Suppose that you are at a park. You see some people throwing a football. When they have finished playing, one of them drinks from a plastic water bottle. She then throws the bottle into a trash can. Do you ever think about where this bottle goes?

A plastic bottle is matter. It has both mass and volume. It does not just "disappear" when it is thrown away. Most likely, the plastic bottle will end up in a landfill. In many areas, however, the bottle could be recycled. Recycling involves taking materials, like those that make up the bottle, and using them to make a new product. In the following pages, you will learn about three common materials that can be recycled. These materials are aluminum, plastic, and glass containers. But before we learn about how to recycle them, let's first learn about how they are made.

Making Aluminum Cans

If you drink from a soft drink can, the can is probably made of aluminum. What do you know about this substance? Aluminum is a silvery-white metal. It is a good conductor of electricity. It is also the most abundant metal in Earth's crust.

Aluminum is an important component of many minerals, such as mica and feldspar. The most important source of aluminum is the mineral known as bauxite. Bauxite can be found in many countries. For example, Jamaica and Australia have large amounts of bauxite. In the United States, bauxite has been mined in Arkansas.

The process that is used to produce pure aluminum was discovered in 1886 by two different people. The two were Charles Martin Hall and Paul Héroult. The two men did not know each other. They lived thousands of miles apart. However, they both made this important discovery at almost the same time.

Before the Hall-Héroult process was discovered, pure aluminum was hard to come by. Today, millions of tons of aluminum are produced every year. However, producing pure aluminum from bauxite requires lots of electricity. This electricity needs to be generated in some way. Sometimes, generating electricity uses up other natural resources.

The Hall-Héroult process uses an electric current to separate pure aluminum from aluminum oxide. The aluminum oxide comes from bauxite. The process occurs at a very high temperature. Therefore, the aluminum is produced as a liquid. Small amounts of other metals can be mixed with pure aluminum to make it stronger and harder.

As the liquid aluminum cools, it is cast into long bars. These bars are called ingots. If the aluminum is to be used for making cans, the ingots are rolled into long sheets. These sheets are rolled up. Then they are sent to the plant that makes the cans. There, specialized machines shape the aluminum sheets into the cans that hold soft drinks.

Aluminum is cast into ingots.

Making Plastic Bottles

As you know, aluminum cans are not the only type of soft drink container. Plastic bottles are also popular containers. Plastics are composed of very long molecules called *polymers*. Plastic polymers contain carbon atoms linked in chainlike structures. This structure allows plastics to be formed into various shapes. Most plastics are made from fossil fuels. Fossil fuels are found in Earth's crust. Coal is an example of a fossil fuel.

Plastics are chemically stable. They do not rot or rust. Stability is useful. But it also means that plastics stay in landfills long after most other materials have decomposed.

There are many different varieties of plastics. Each type of plastic is composed of specific molecules that are linked together in a particular way. For example, polyvinyl chloride (PVC) contains chlorine atoms attached to the "backbone" of carbon atoms. PVC has many uses. The rigid form of PVC is used in home products such as piping and siding for houses. It is also used in products such as shampoo bottles and medical tubing. A softened form of PVC can be made by adding other chemicals. This type of PVC is also used in many products, including shower curtains and raincoats.

One type of plastic commonly used to hold drinks is polyethylene terephthalate (PET). This type of plastic is often used to package soft drinks. The PET used in plastic bottles is strong but can still bend somewhat. And, because it is clear, you can see the liquid inside. PET has a wide range of uses. For example, PET is used in many car parts. It can also be used to make fabrics, carpets, and pillow filling. Remarkably, many things can be made from just this one type of plastic!

At the grocery store, you have probably noticed that plastic milk containers look different from plastic soft drink containers. This is not surprising. Most plastic milk containers are made of high-density polyethylene (HDPE). HDPE in milk containers is strong and stiff. It is also translucent. This means that it allows light to pass through it only partially. This is a useful property because sceintists think that too much light can affect milk's taste and nutritional value. Pigments can be added to HDPE and other plastics. These pigments give the plastic different colors. Plastic grocery bags made of HDPE come in a variety of colors.

Plastic bottles can be produced by a process called blow molding. One method of blow molding involves placing a molten tube of plastic into a mold. Air is then blown into the tube. This causes it to expand and take the shape of the mold. The process resembles blowing up a balloon. When the plastic cools, the mold is opened, and the plastic container is removed. Like aluminum cans, plastic bottles are produced quickly by machines. Milk bottles and soft drink bottles can be produced by blow molding.

Blow molding forms plastic containers.

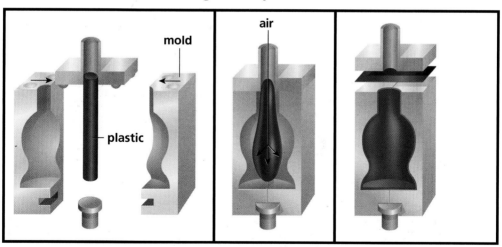

Making Glass Bottles

Like aluminum and plastic, glass can be used to hold drinks. Glass has unusual properties. At normal temperatures, it is hard and brittle. However, it is not classified as either a solid or a liquid because its particles are not arranged in an orderly way. For this reason, glass is called *amorphous,* which means "without shape."

Glass contains a large amount of silicon dioxide, or silica. Silica is obtained from sand. Glass is produced by heating silica to a high temperature, usually with other raw materials, until a liquid forms. Liquid glass has properties like those of plastic. It can be shaped in different ways. It can be pressed into molds, blown into hollow shapes, and made into flat sheets.

There are several different types of glass. The properties of a particular type of glass depend on the substances it contains. Glass containing lead oxide is heavy. It bends light more than most other types of glass. Crystal table glass is an example of glass that contains lead oxide. Another type of glass, called borosilicate glass, contains the compound boric acid. Borosilicate glass is durable. It can withstand higher temperatures than some other types of glass. It is often used for cooking and lab work. Color can be added to glass by mixing in metallic compounds.

Lead oxide in crystal glass increases the bending of light.

You have probably seen glass in many different shapes. How can the glass take so many shapes? One way to shape glass is by blowing it.

You may have seen glassblowers at work. The blower gathers a ball of molten glass onto the end of a long iron blowpipe. He or she then blows a glass bubble. The bubble can then be rolled on a slab and shaped with tools. More blowing and shaping can follow. The glass is reheated as needed. When the glass is formed into the desired shape, it is removed from the blowpipe. The opening is then shaped. Glassware can be decorated by engraving, painting, etching with acid, or other methods.

Glass blowing is usually used to make pieces of art. Today, though, most hollow glass containers are made by machine. One way machines make glass bottles involves a two-part process. First, a blast of air sets a mound of glass in the bottom of an upside-down mold. This forms the top part of the bottle. The half-formed bottle is then placed into a second mold. Another air blast blows it into its final shape. One advantage of glass containers is that they do not affect how a drink tastes. On the other hand, glass breaks easily. It is also thicker and heavier than aluminum or plastic containers.

Glassblowers make artistic glasswork.

Why Recycle?

Now you know how different types of containers are formed. But how do these containers affect the environment? Throwing away containers takes up space in landfills. And, as you have learned, it takes a long time for plastics to break down, if they ever do. What can be done about these containers? Do you remember the plastic bottle that was thrown away at the park? That bottle could probably have been recycled. To recycle a bottle involves reusing the materials that make up the bottle to make new products.

Many communities have recycling programs. Among the commonly recycled materials are aluminum, plastic, glass, and paper. Families place recyclable materials in special containers to be picked up. Recyclables can also be taken to drop-off centers. The recycled items are taken to a materials recovery facility. They can then be made into new products. You may ask, though, why recycle when it is easier just to throw things away?

Recyclables can be left at drop-off locations.

One important reason to recycle is that it saves natural resources. Recycling reduces the amount of original material needed. For example, recycled aluminum is made into new cans. This reduces the amount of mined bauxite needed. Similarly, recycled plastic can be used to make new products. This reduces the amount of fossil fuels needed to make new plastics. Recycling also conserves natural resources by saving energy. For example, producing an aluminum can from recycled metal requires less than five percent of the energy needed to make the same can from raw materials. Recycling glass also saves energy. Why is saving energy important? Fossil fuels, such as coal and oil, are important sources of energy. Fossil fuels are a limited resource. They can be used up. Using less energy conserves these fossil fuels.

Another advantage to recycling is that it reduces pollution. For example, recycling aluminum reduces the need to mine and process aluminum ore. Both of these activities create pollution. Also, by reducing the consumption of fossil fuels, recycling lowers the amount of carbon dioxide released into the atmosphere. Carbon dioxide is a gas that is believed to cause global warming.

Recycling reduces the need for landfill space.

Finally, recycling reduces the need for landfill space. Most solid waste in the United States is sent to landfills or dumps. These landfills eventually fill up. New landfill sites can be difficult to find because people living in an area often do not want a landfill near their homes.

One way to limit the need for landfills is to recycle whatever you can. Currently, the United States recycles only about 28 percent of its waste. There is plenty of room for improvement. In 2000, overall recycling of glass was about 23 percent. The estimated recycling rate for plastic bottles is much higher. In 2000, the estimated rate for PET soft drink bottles was 34.9 percent. Also in 2000, the estimated recycling rate for HDPE containers (such as milk jugs) was 30.4 percent. The news for aluminum is just as exciting. In 2001, 40 percent of aluminum in containers and packaging was recycled!

How Materials Are Recycled

Aluminum cans are collected for recycling. They are shipped to a reclamation plant. There the cans are shredded and heated. The aluminum is then melted in a furnace and formed into large ingots. These ingots can be used to make new cans or other aluminum products.

Aluminum cans are recycled.

Glass comes in different varieties and colors. This makes it somewhat more difficult to recycle than aluminum. Container glass is most commonly recycled. Its composition is different from other types of glass, such as window glass. Therefore, container glass is not mixed with other types. Recycled glass containers are separated according to whether they are clear, brown, green, or another color. The separated glass is melted so that it can be made into new containers.

Recycling plastics is also not easy because so many different types of plastics are used in containers. Plastic waste must be sorted by the type of plastic that makes up each item.

The sorting is made easier by identification codes. Each type of plastic has a particular number code. For example, the number 1 is used for polyethylene terephthalate (code abbreviation PETE, but more commonly known as PET). The number 2 refers to high-density polyethylene (HDPE). There are seven different plastic codes. The table lists each of the seven types and examples of their use.

Recyclable Plastics		
Code Number and Abbreviation	**Name of Plastic**	**Examples**
1 PETE	Polyethylene terephthalate	Soda and water bottles, food containers
2 HDPE	High-density polyethylene	Milk jugs, detergent bottles, grocery bags
3 V	Polyvinyl chloride	Plastic pipes, shower curtains
4 LDPE	Low-density polyethylene	Bread bags, squeezable bottles
5 PP	Polypropylene	Car battery casings, long underwear
6 PS	Polystyrene	Coffee cups, egg cartons
7 Other	Other	Various

After sorting, plastics are usually cleaned and shredded into flakes. These flakes are melted into pellets. The plastic pellets can be remelted to form new products. Recycled plastics are not normally used to make food or beverage containers. For example, PET can be recycled into carpets. HDPE can be used for products such as detergent bottles and trash cans.

Conclusion

At the beginning of this reader you were at a park, watching someone throw away a plastic bottle. Perhaps now you have an idea of how that bottle might be put to good use. Recycling is a good choice. If there is a recycling program in your area, check the guidelines to see which materials are collected and which are not. And remember that recycling is not the only way to conserve resources and protect the environment.

You can also help by reducing the amount of material you use. Reuse items whenever possible. For example, you can write on both sides of a sheet of paper instead of only on one side. You can also rinse out plastic jars and use them to hold items such as pennies or paper clips. Perhaps your family has reusable items that you no longer need. These items may include used furniture or clothing that you have outgrown. If so, you might give these items away instead of throwing them out. By making better choices about the products you use and how you dispose of them, you can help preserve the environment and natural resources.